THE ENERGY OF LOVE TRIUMPHS ALL WORRY, ALL
FEAR, ALL QUESTIONING, ALL TIME, ALL SPACE.
IT HAS NO BEGINNING, AND NO END.

Love lives forevermore

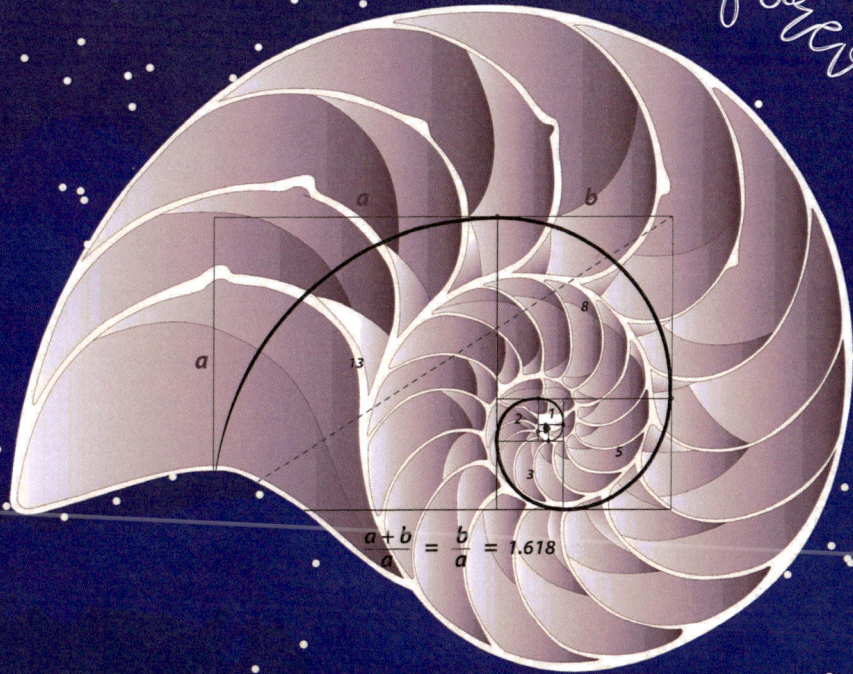

$$\frac{a+b}{a} = \frac{b}{a} = 1.618$$

Interior and cover design: Cassandra Neece

National Library of Australia Catalogue-in-Publication data:
Quantum Love/Adrea L. Peters and KP Weaver

ISBN: (hc) 978-0-6450521-5-2
ISBN: (e)   978-0-6450521-6-9

QUANTUM
LOVE
factoring in Infinity

ADREA L. PETERS & K P WEAVER

MAY YOU BUILD YOUR ENERGY
and your love
FROM WITHIN
and watch, in utter awe,
AS MORE ENERGY AND MORE LOVE
arrive
TIME AND TIME AGAIN.

TABLE OF CONTENTS

Welcome

There is no force I adore more than love. It wields the best kind of power. It handles healing with ease. It crushes arguments. It dominates happy people. Love cares about everyone. Love listens. Love hears. Love sees. Love is who I want to be in every minute of every day.

Karen feels the exact same, which is why I welcomed her to create on these pages with me. I'm beyond honored to get the opportunity to write each line of this book with my dear friend. We spend quite a lot time communicating on a daily basis and what I love most is that we always focus through the lens of Love and Gratitude. We discuss the to-do list of various publishing projects, of course, but we make time to share our latest favorite discoveries, our feelings about nearly everything, and always what we are *loving* about life.

We operate with a gentle, subtle undertone of pure love for everything. Good *and* bad. Right *and* wrong. Today *and* tomorrow. And that is the absolute heart of the Quantum collection. Possibilities abounding. Limitation-free. Living a life filled with love, elation and unending harmony.

The energy of these pages is the energy of love. All frequencies. All waves. All love. Kindred. Familial. Romantic. Natural. And Infinite.

May you feel the love in every moment you are within it.

All our love,

Adrea & Karen

1

Unconditional

The beauty of unquestionable love given freely and reciprocated equally is a wholehearted experience. Unconditional love is like stepping into a dream that comes to life without restriction. How do we know when the love we are giving is unconditional? We have no expectation of love being returned. And when the love we are given is unconditional, it gifts us the freedom to be exactly who we are at all times. Unconditional love is a love that allows us to be curious, to explore ourselves, and to figure out the incomparable power of love.

I love myself unconditionally.

Harmonize

When we are singing and another comes into the song, we harmonize by retaining our own voice. We can be distinct and in sync with each other and create magic. Love relies on us being in harmony with ourselves, including knowing what matters and what we need to let go of. It is imperative to know what you value in order for another to value you. Love yourself enough to make time to write your song before you harmonize with another.

I harmonize with myself before
I harmonize with others.

2

3

Growth

Your desires and dreams are the seeds of your life. We plant lots of seeds in our lifetime. The ones that we choose to nourish take root and grow until they blossom. Growth has a thirst to be nurtured. When we love, our capacity to grow is increased in the best possible ways. Love lets us grow beyond all limitation. A mind infused with love inspires endless growth for us all. No fear, no worry, everything is growth.

I take an active role in my growth.

Oneself

To love oneself is to receive an unending stream of love. Every day love is guaranteed to shower you. May we all begin a beautiful love affair with ourselves. May we gift ourselves a sweet love letter as often as we need to remind ourselves exactly how lovable we are. May we gift ourselves what we need... a little quiet time, a bouquet of flowers, a new book, a moment with nature. We cannot fully love another until we fully love ourselves. What was the last thing you did to show yourself that you love yourself?

I am in a beautiful love affair with myself.

4

5

Let

You don't need permission to love. It doesn't matter if the person you love loves you. Love them anyway. We are the doorkeepers of the love we let into our lives. When we let love lead, we can trust that love is what we will receive. Love makes the world go round. May we let love navigate.

I let love navigate my life.

Share

Love is not something to be kept within. It is meant to be shared and create a ripple effect throughout our world. Love unites. Love conquers all. When we chose to share love instead of hate, fear subsides. Ask yourself: what would love do? Love would love.

It is safe to share my love.

6

7

Gratitude

There is no greater power than love-fueled gratitude. Multiple studies have shown that keeping a list of what and who you are grateful for will improve your outlook on life exponentially. A simple "thank you" can change the world. Each day, may you show yourself gratitude by having a mindful moment to feel grateful. Gratitude is the path to the highest vibration there is: love. If you want to see results in your life, start and end each day with gratitude. (And feel free to sprinkle some gratitude into as much of the middle of your day as you like.)

Gratitude is my guide.

Energy

The energy of love gives us endless vitality and stamina. Love is the fuel of the heart, mind and soul. Love really is *all you need*. When we love, we become a beacon that opportunities gravitate to. When love is our anchor, we are connected to all that is. Love never taps out. The energy of love is always there so we know that no matter what, we are always fueled.

I am the energy of love.

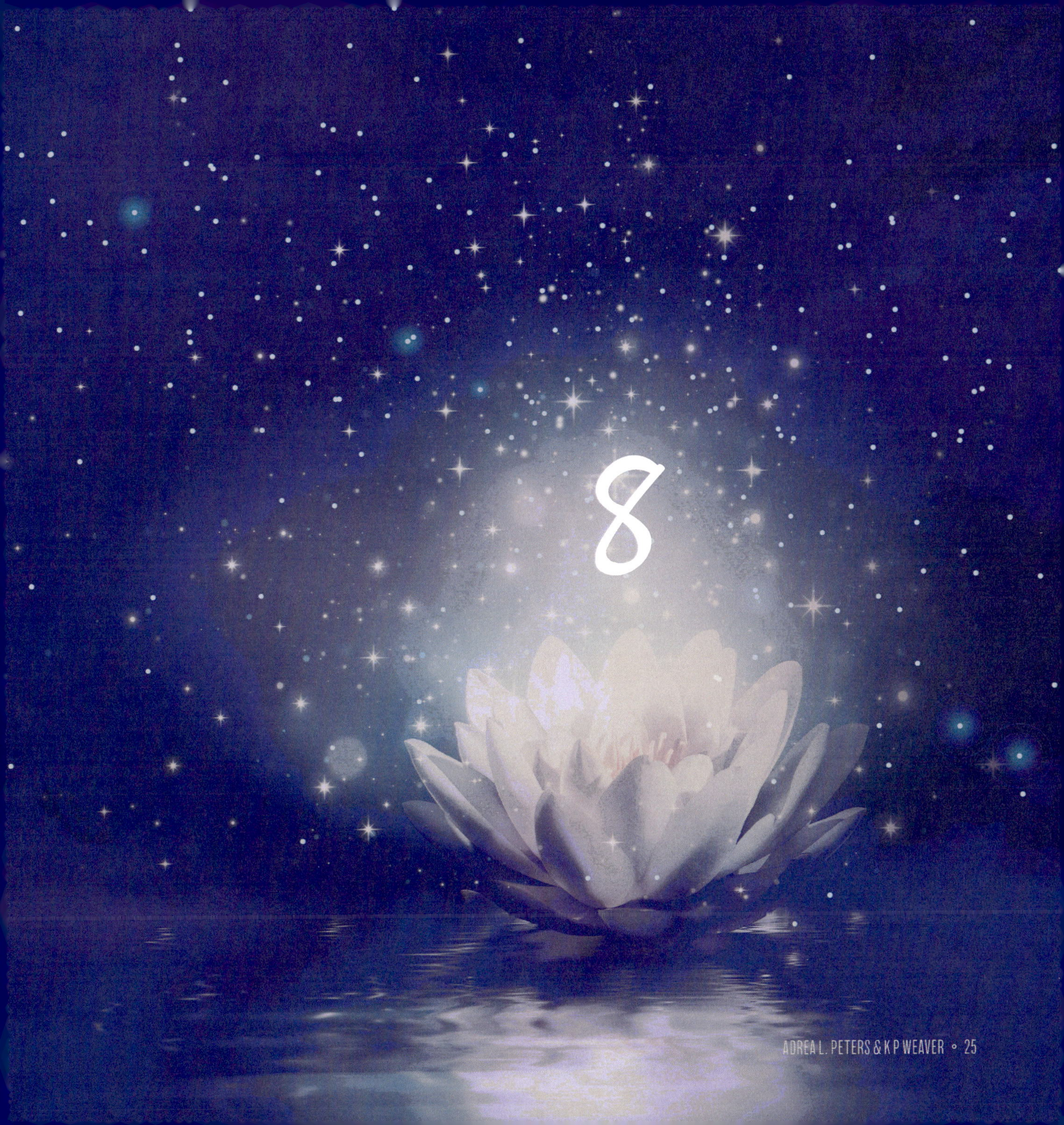

8

9

Connect

Genuine connections are built on the foundation of love. When we show up with love, we connect. Our connections are a reflection of where our love is. They are gauges for how we are treating ourselves. Connections are magnetic. Sometimes we have to break the connection, learn to love ourselves more, then let love determine how we will connect again, or not. Love holds no grudges. It opens new doors.

I expect to connect with those who love.

Infinity

Love is infinite. It has no beginning and never ends. Love is. Love always was. There is no limit to the love we can give or receive. Love doesn't dilute. It never runs out. Close your eyes and imagine the night sky surrounding you. It goes and goes and goes. Look right, more. Look left, more. Always more love.

I am infinite love.

10

11

tender

You don't need to be so tough. Being tender is what makes us stronger. It's easy to attack or gossip or put ourselves down out of habit. Making tenderness a priority could be the one thing you do that changes everything.

I nurture the tender me.

Wild

Is there anything better than unabandoned love? To be wild in our love, we enjoy loving. Love loves to play and to dance and be free. When we love life, it loves us back. Wild doesn't mean irresponsible, it means curious, and open-hearted and eager to align with others. Being wild is how we find our truest potential. You have to try things out, crash and burn, then watch love pick you up and dust you off to go on another wild ride.

I love the wild inside me.

12

13

Elation

Elation is happiness on steroids! It is a state of being fresh and vibrant and blooming with love. When we are elated, loves oozes from us and everyone's interested in who we are and what we're up to. Find elation and watch the love pour in. What is it that makes you vibrate? Do that. It will spark all sorts of elation in your life.

I live in the elation zone.

Friendship

Friends are the family we choose. When we decide who we bond with, we give ourselves the gift of the deepest kind of love. A true friend is one who wants you to be more yourself than you know how to be. A true friend is your greatest champion. Friendship literally saves lives every moment of every day. Friendship is the absolute best part of being human. It is consciousness on fire.

Friendship is where I can be my truest me.

14

15

Heart

Your heart is your compass. It does not need be shielded. It is strong all on its own. After all, it's connected to infinite love. May we have the courage to know that love always has our back. A heart never breaks. It always beats to the rhythm of love. May you appreciate the steady flow of love going straight into your heart.

I touch my heart to honor
how loved I am.

family

This is where we all start. We are born into our family. It may not last more than a moment. It may span endless lifetimes. May we welcome the love of family. Families give us the fortitude to find our footing and push our buttons and boundaries. Family holds us and nurtures us. Family goes out of the way to make us whole. Family is always a source of love.

I let my family love me.

16

17

Certainty

Love is certainty. Love is predictable and consistent. Loves loves. Love says, "I have no doubt." Love is a constant. It will wait for you to travel through uncertainty. May you be certain that you can always rely on love.

I know my certainty comes
from love.

Participate

In order for love to flourish, it requires participation from everyone involved. To fully be in a loving partnership, romantic or platonic, we engage, we play, we co-operate, and we collaborate. Love is not one-sided. It's multi-dimensional. Love wants us to get in the game. To love someone is to commit to showing up and participating each and every day.

I participate in all my partnerships.

18

19

Satisfaction

There is a release that love gives us that is the ultimate satisfaction. When we give love, we receive love. Loving is perpetually satisfying. There is so much pleasure in loving. Love is everywhere. In people, in animals, in words and art, in all the beauty of our planet. May we relish in the satisfaction of a love-filled life.

When I love, I am satisfied.

Radiant

To possess love is to be radiant. It is a beam that cannot be ignored. Everyone and everything lights up when the radiance of love projects itself. May we be incandescent sources of love in the world. The light of love never dims. It shines and shines and shines evermore.

I am radiant.

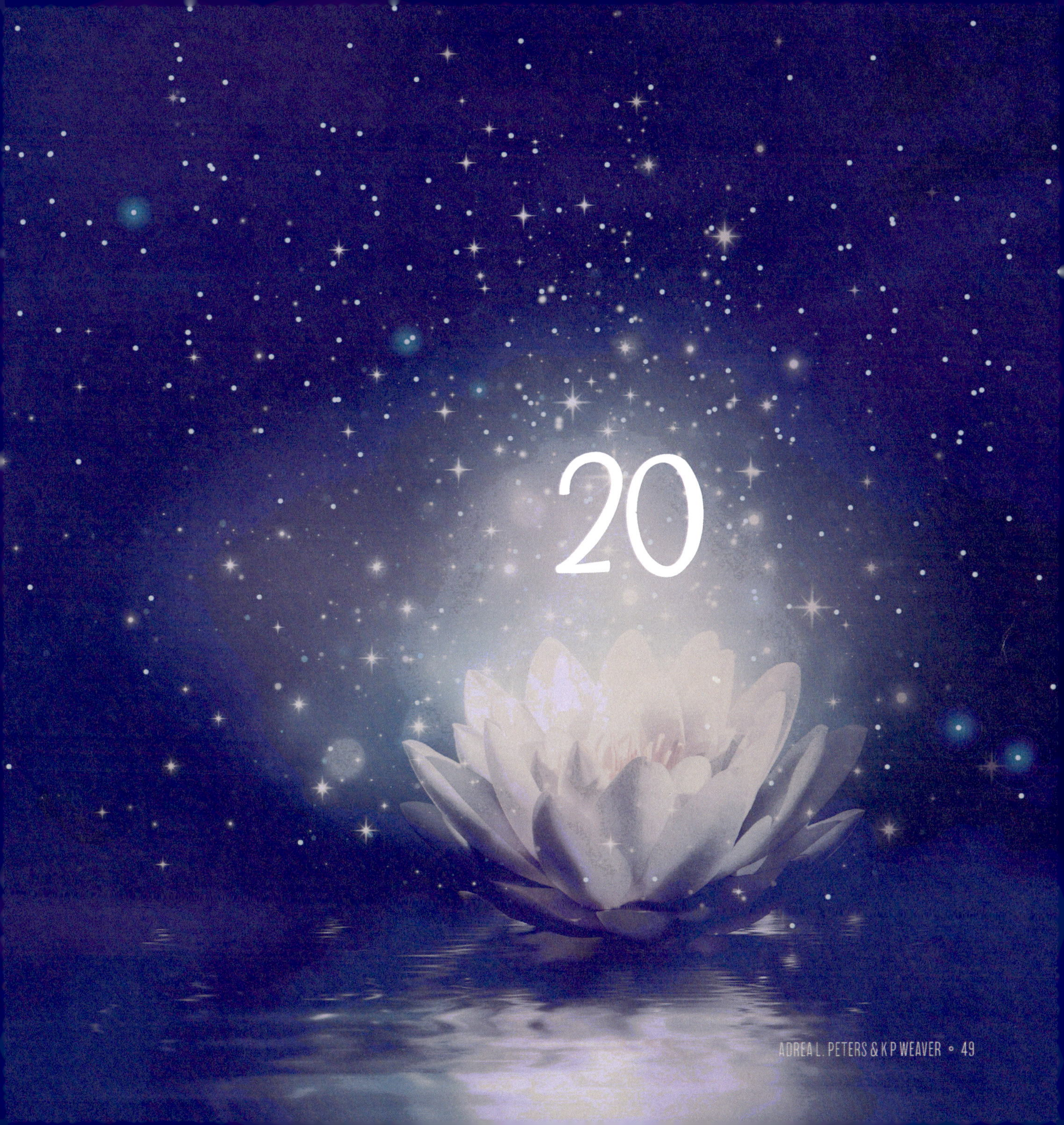

20

21

Cherish

Cherished love is celebrated love. When we cherish another, we love them wholly. When we cherish ourselves, we hold ourselves in the highest regard. To cherish is to admire, to appreciate, to think the world of, to treasure and hold dear. May we all feel the full effect of cherished love.

I cherish my life.

Serendipitous

The ultimate payoff of a life filled with love, is a life filled with serendipitous events. Serendipitous means discovered by chance in a happy and beneficial way. Love enables coincidence to be reality. Love sorts everything out for you. You and the Universe are nodding and winking at each other all day long. May you have a life of endless serendipity forevermore.

I bounce from one good thing to the next in my incredibly serendipitous life.

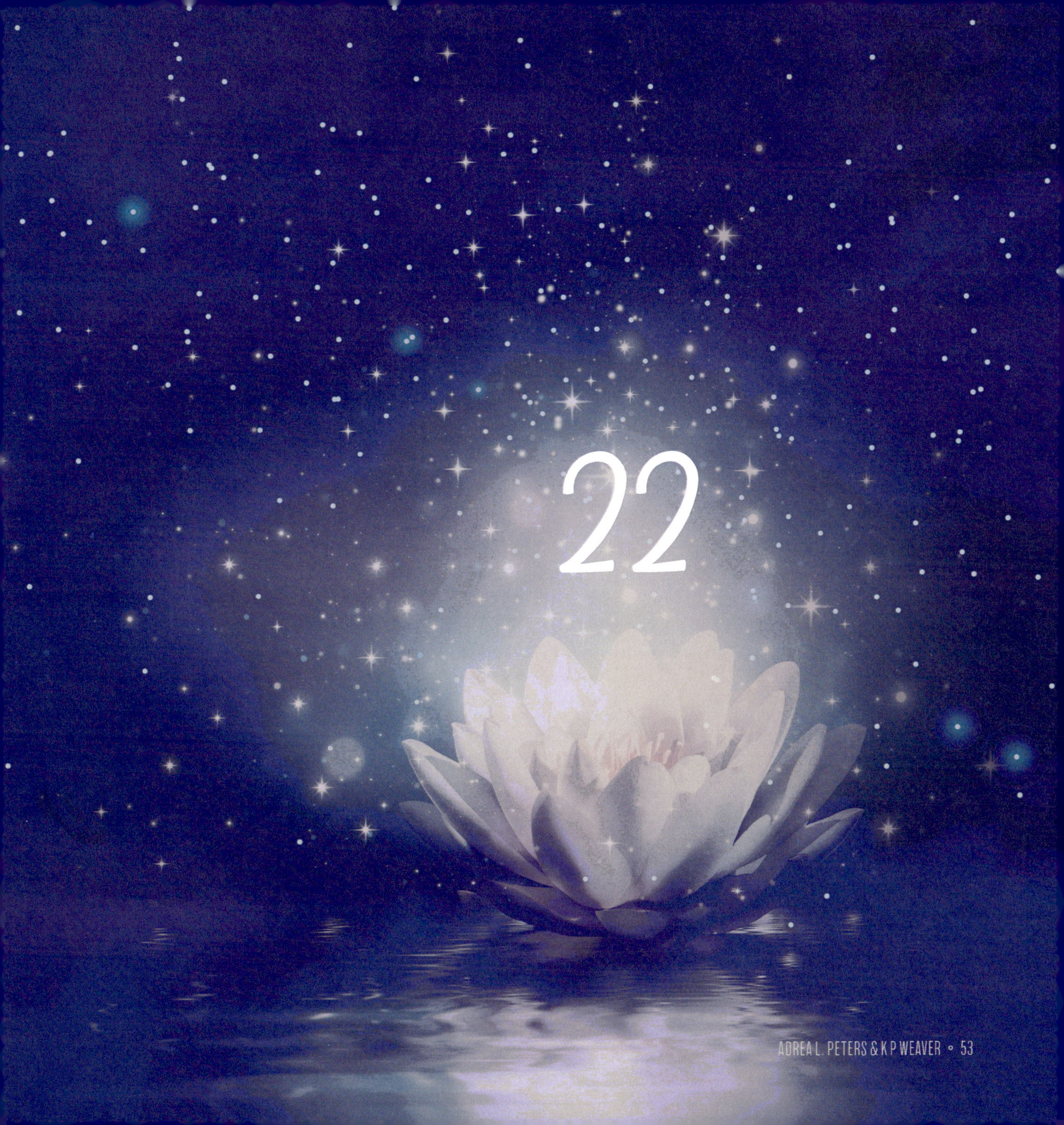
22

ACKNOWLEDGEMENTS

Our dear readers, we love you! You are the reason we create. It is such a pleasure to deliver this gift to you. Thank you for being a part of the most beloved part of our lives.

Cass. We love working with you! You are love! Thank you for being so divine to collaborate with! We cherish and love you.

The KMD Books team. It is such a pleasure to work with all of you. We feel held and loved and nurtured endlessly. Thank you.

From Adrea to her family and friends, you all taught me love and I thank you for it. I love you. I love you. I love you.

From Karen. To my mum and dad who gifted me the honour of knowing what true unconditional love was from the moment I was a twinkle. To my children, friends and readers thank you for receiving my love with open arms.

ABOUT ADREA

Adrea is a bestselling author-shareholder with KMD Books. She trained as a journalist, novelist and screenwriter. She earned a Bachelor of Science in Journalism from the University of Colorado at Boulder and Master of Arts in Fiction Writing from Seton Hill University.

She is the author of the *Becoming Truitt Skye* series and the creator of the *Quantum Collection*.

Her bestselling book, *Quantum Thinkin*φ, released in September 2020. She co-authored, *When I Go Outside, I Go Inside* with longtime writing partner and dear friend, Teffanie Thompson, which released in late 2020. Her book and companion journal, *The Science of Story (SoS)* were written for all of us to understand and embrace our unique stories.

She lives amongst the trees in Vermont with her boys, Skye and Fig. Please visit her website at www.adreapeters.com to connect and learn more about Adrea and her books.

Visit www.adreapeters.com.

OTHER TITLES BY ADREA

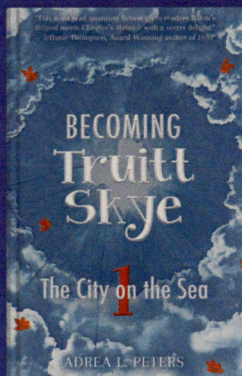

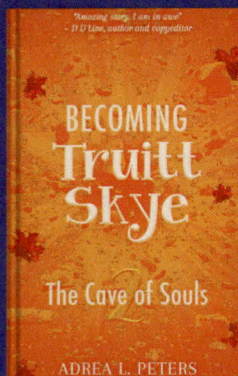

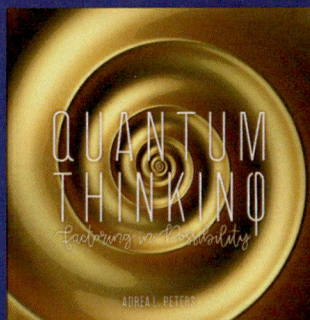

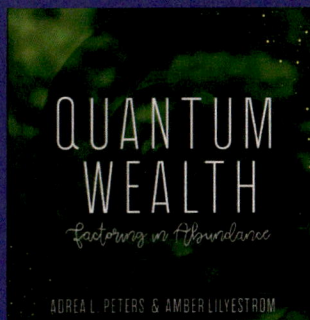

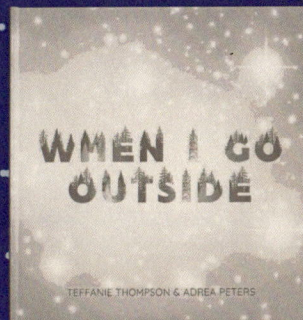

BECOMING **Truitt Skye** — The City on the Sea 1 — ADREA L. PETERS

BECOMING **Truitt Skye** — The Cave of Souls 2 — ADREA L. PETERS

The **Science of Story** — Mastering Your Nature — ADREA L. PETERS

QUANTUM THINKING — *Factoring in Probability* — ADREA L. PETERS

QUANTUM WEALTH — *Factoring in Abundance* — ADREA L. PETERS & AMBER LILYESTROM

WHEN I GO OUTSIDE — TEFFANIE THOMPSON & ADREA PETERS

ABOUT KAREN

Karen P. Weaver is an award-winning publisher, author, and advanced Law of Attraction practitioner.

Author of numerous books across many genres including her non-fiction Alchemy of Life Magic series and fictional novels including]The Enlightenment Series, and has fun writing children's books under Mamma Macs.

She chooses to lead the way in her authorship by sharing her life philosophies through her writing. Karen is also a sought-after speaker who shares her knowledge and wisdom on building publishing empires, establishing yourself as a successful author-publisher, and book writing.

Having built a highly successful publishing business from scratch, signing major authors, writing over 30 books and establishing her own credible brand in the market, Karen has developed strategies and techniques based on tapping into the power of knowing to create your dreams. Karen is a gifted teacher who inspires others to make magic happen in their lives through utilising her power of knowing strategies. She mentors multiple new authors and publishers through her Everything Publishing Academy.

Visit www.kpweaver.com.

OTHER TITLES BY KAREN

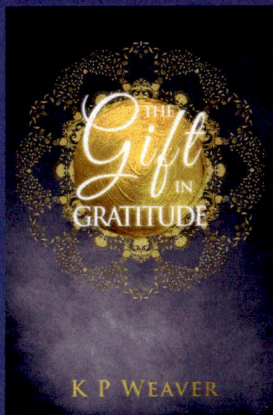

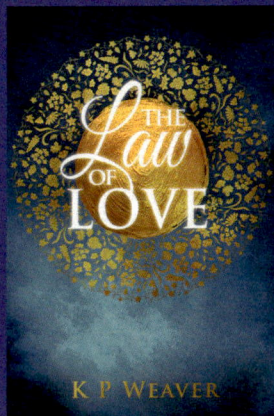

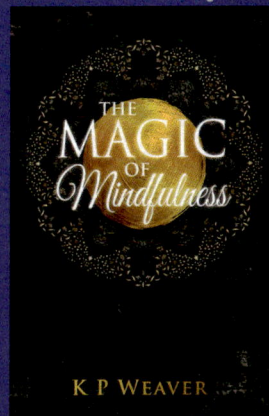

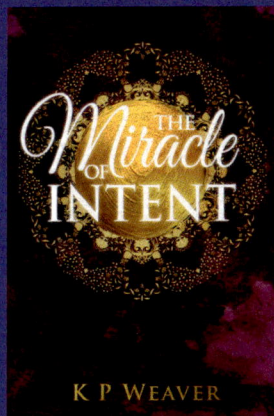

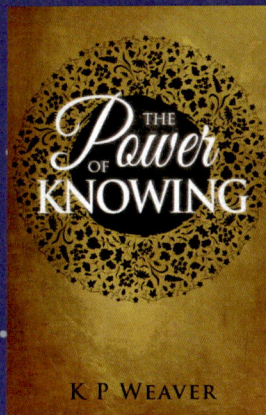

THE Gift IN GRATITUDE

K P Weaver

THE Law OF LOVE

K P Weaver

THE MAGIC OF Mindfulness

K P Weaver

THE Miracle OF INTENT

K P Weaver

THE Power OF KNOWING

K P Weaver

www.ingramcontent.com/pod-product-compliance
Lightning Source LLC
Chambersburg PA
CBRC091800090426
42811CB00021B/1898